New Resources
for
Individual Psychological Diagnosis

Between Cognitions and Emotions
The Research Context

Version 3.0

Jan Sterenborg
2018

PREFACE

In "New resources for Individual Psychological Diagnosis" a diagnostic instrument is presented. The instrument can be used both in the field of cognition as in the field of emotion and is meant to be for daily practice. The cognitive approach provides verifiable data and the approach to emotions provides descriptive data in a subtle interplay of conscious and unconscious processes. The unconscious is seen as the absolute reliable experience basis in every human. The instrument is designed to let the unconscious speak in a conscious way.

This book is dived in two main chapters[1]:
1. Between Cognitions and Emotions, in which the new diagnostic tool is described as an elaboration of the work of Gé Calis.
2. The Research Context, in which an overview and status of the different research approaches is given. This chapter can be seen as the metho-dological basis for scientific psychological research.

---

[1] At the end of this paper a schematic overview of the approach was added and an introduction to the Symmary, the technical literature and the Summary itself.

Dedicated to Gé Calis

Between Cognitions and Emotions

A diagnostic instrument in the making

Introduction

When Sigmund Schlomo Freud postulated that unconscious forces were active in every human being, psychology as a science was born.

From there on numerous developments were started which more and more took the form of scientific research and experimentation.

Can we prove what we claim? Can we establish key facts concerning man? This line of development reached the point that the unconscious was totally abandoned as unreliable.

But what was unreliable: the interpretation of the unconscious or the unconscious itself?

In this paper the unconscious is treated as the absolute reliable source of knowledge and experience in every human being and the question to be answered becomes:

How can we let the unconscious speak in a conscious way?

A circle is completed: the unconscious and the conscious come together in a subtle interplay. The approach described in this paper provides us information concerning this individual human being.

The method will be called: Individual Psychological Diagnosis (IPD).

Man stays a mystery

We will know more and more about people, but there will always be, paradoxically, more things we do not know. And even though the manual for psychiatry (DSM) becomes three times as thick, it will not solve things substantially (see first part of Stranger at Killknock).

But let us be modest in our approach to man. As the rest of this paper will show, we cannot say much scientifically about man in his mental functioning. Scientific in the sense that we can prove what we claim. There is only one field where this is possible and that is in the field of concepts or cognitions: has a person a distinctive concept to his or her disposal?

And how are concepts related to each other within this person?

In case of learning processes this is very important to know, but whether someone is schizophrenic or suffers from another mental disease, we can find clues, but as hard facts these clues are never to be found.

The same phenomenon seen from the outside can have many different inner causes.

There are of course people who can say something significant about a person on the basis of their experience and insight, but what they have to say is entirely for their own account.

In seventies of the last century, Nijmegen University was a kind of museum of all the methods that were developed in psychological research. There I met Dr. Gé Calis who was a senior lecturer at the Department of experimental psychology and his research field was the visual perception.

In 1974 he obtained his doctorate with a thesis entitled:
"*At first sight. Immediate perception and facial recognition.*"

Calis approached things in a way that when an understanding of the human being in general is true, this also should to be true for the individual. So no throwing on piles. This appealed very much to me because I was looking for a research method that could tell me something about the individual.

What I did not know then but know now is that this road was not an easy road. A dragon with twenty heads had to be slain.

In the period 1977-1997 this dragon was slain.

Perception
Although the presented research focuses on visual perception a more general question goes first: what is perception?
We see something when we are not disabled, we hear something, we smell something and we think something, all forms of perception.
"Something" is being perceived: this something distinguishes itself from other things.
We see a car, although much more is to be seen, houses, trees, other cars, people, the air and so on. Our attention is with the car and we isolate something in the field of perception and name that a car and when we know more about cars it becomes a BMW or a Humber.
Perception is so direct and immediate that we assume that someone else with normal vision capabilities also sees the thing I am talking about. This immediacy is so strong that we even assume that we both see exactly the same thing.

Do you see that car?
Yes nice one, a BMW! My father has one.

And so the idea arises that there is an independent reality out there, apart from the observer: a reality that is the same for everyone. And that we can pin point that reality and worse we can impose that reality on a person, this is it and you have to see that too!

Too bad, it is yours and mine reality and the problem becomes to establish whether what you see, do I see that too? The birth of science. Or birth of a scientific attitude and also bound to the person in question and not something that functions as an independent, anonymous something outside people.

How does this scientific attitude look like?

The basis of a scientific attitude is: *repeatability*.

In other words: I think I saw something and now I describe it in such a way that another person can see that too. When more and more other people confirm my findings, then this finding gains credibility.

A carpenter knows this thing: he has to produce a table and four identical chairs. In the world of parts it is not different: when something is broken I want to replace the broken part to fix it.

In the field of scientific publications it's a mess lately, too many scandals are coming up: one publication is enough to become a hero in the field, decisions are made on one research result and that is very alarming.

Only research that is repeated adds to the credibility of the original research.

How do we get a grip on (visual) perception?
Not a simple matter because when we open our eyes and there are no disabilities then directly and immediately there is a world visible. A world with all sorts of concrete things.
This immediacy silences all questions.
The questions come back if we want to build a device that can see.
For example, we want to make a device that can read handwritten text. This device would be very helpful in, for example, the sorting of mail.
We helped a bit by simplifying the problem, and entered postal codes which are easier to read than only the hand-written text.
We need to know (increasingly make explicit) how perception works, otherwise it remains a flawed and clumsy device.

These developments are in full swing, just think of robots. Our cars will be equipped in the future with observing systems. We need to drive our cars no longer. In other areas, we find the same developments. In Belgium in a home for the elderly, robots take over the care and the reason that had been put out by the media: there are not enough people to be found who can care for the elderly. And that of course is a fallacy: there are certainly enough people who want to care for others, only robots are cheaper, a purely economic motive with the risk of social cohesion disappearing completely. In South Korea, a robot runs through the prison corridors to check the cells. In China there is a restaurant with robots to operate. The robots are programmed so that they only react kindly to customers.

Who is in charge? Man or robots?
The difference between a man and a robot will be minimal. Man is already robotised by bureaucracy and regulations. Humanity is at stake.
Goethe (1797) wrote in his sorcerer's apprentice: "Die ich rief, die Geister, werd' ich nun nicht loss." "From the spirits that I called, Sir deliver me!".

The approach of Gé Calis
When we read everything what has been published about perception, we get dizzy. A jungle of publications and ideas and to confront that and deliver something substantial is not easy. Calis had the courage to do so and in my opinion he gives the right direction. On the basis of his study about visual perception and his own research Calis concludes that perception is hierarchically organized before *something becomes something* as an identifiable object.

The hierarchical process goes from general to specific and in the end the "thing" can be named.

An example:
Walking in town hoping to meet a friend to drink coffee with. Let's have a look... isn't that.... No it isn't. Peter wears glasses and this person not although his hair and contour looks like Peter and over there is that Marie? Approaching her she turns her head to the right and we see a total different Marie. Our Marie has no freckles! In the end: "Yes," there is John "Hallo John, fancy a cup of coffee?"

When we look at the above scene for the person who is looking for a friend and is confronted in his search with numerous people he sees coming towards him, internally there has to be a reduction from possible people to a known friend. Think of a robot walking in the streets and that is instructed to recognize John. How should that work? Has the robot a picture of John? And compares the robot all the people he sees with this picture? Imagine the time that is needed to do so and further more perhaps it is not a recent photograph of John the robot has, so it becomes nearly impossible to do so. Some intelligence must be brought in to do so.

A great help would be if the robot could establish whether the person he sees is a man or a woman, this would save lots of time. And whether the person he sees is old or young or wears glasses, has brown hair, is ten feet tall and so on. In the end after all these measurements 1 person is left and we have a positive confirmation: this is certainly John (later on it turns out that John has an identical twin brother Peter). Remember that as a perceiver this process establishes the identification in a moment, we are not aware of this process.

This all looks a bit artificial but looking at the sorting machine that has to read handwritten text on mail: in the end the machine has to identify 24 lower case letters or 24 upper case letters then 10 numbers (left aside the roman or other indications of floor). Is the letter written in English or Arabic? Is this the front or the back of the letter? Does the machine look at the front upside down? And above all these things perhaps the person who wrote the address has made a mistake. For a normal perceiver naturally and simple acts for a machine not.

Some intelligence (a program) has to be active within the machine. Features have to be established and from there on decisions have to be made and all of this in very little time.

There is necessity in the order of steps, a logic: you have to first capture the position of a head before you or the system can establish more detail of the face: position of eyes, mouth, ears, nose and so on.
As long as the position of a face is not known, all further actions to find new features within the face, become fruitless. And certainly when the time to act is limited. But a human being has other means to establish identity of another person e.g. the voice of that person and specific movement patterns.

    Who is this ?---A---B-----C-----------! Name identification
    >> *time*

Let's look at the possible steps in an identification process: the number of steps we don't know, it could be one step when we guess the name of a person as the right name or more steps when we actually perceive: this means in interaction with our world.

The question to our scientific researcher is to establish whether the postulated steps are really active in the person to be investigated. And then answer the question whether for more people this is the case.

The question: "Who is this?" seems simple but when the person to be investigated doesn't understand this question we can stop already.

How can we establish whether the perceiver has understood our question? We can determine whether our perceiver has understood the question by presenting this perceiver a couple of pictures and ask him or her: Who is this? And then establish whether this perceiver gives *repeatedly* the same answer. Or when the perceiver cannot speak, he or she presses the same button belonging to the shown picture over and over again.

Calis 'research, published in Acta Psychologica, can be summarized as follows:

Who is this ? --- face-Position------Spectacles ------ ! Name Identification
>> *time*

The perceiver starts with the question: Who is this? Then a short movie is presented, a movie containing only two frames. Two pictures after each other both exposed very short. The position of the heads on the two pictures is varied. Looking to the left and looking to the right. Half of the persons to be identified wearing glasses.

The response set for the perceiver consisted of six persons from which three wear glasses.

The line of reasoning is that if the perceiver establishes on the first presented picture the position of the head, he or she will profit from this fact when identifying the second picture. And secondly: when he or she establishes also on the first picture whether the presented person wears spectacles he will have more benefit when the person on the second picture has the same position and wears spectacles too.

The research results confirm this line of reasoning. A real scientific achievement in my opinion and the first real facts in the history of psychological research.

Modifications

There are lots of reasons why experiments are not repeated. Perhaps the specialized equipment is not at hand. Timing is not right? Nowadays one can repeat the experiment on a home computer or laptop but then one needs some computer programs to do the job. Who is reading scientific articles?

To me an important role to go on with Calis approach was interpretability and the practical use of the approach. That led me to a simplified approach. These simplifications have led to repeating the experiment of Calis with good results and an elaboration of the application.

- Calis used in his experiment for the first and second picture the same set of pictures.
- The total exposure time of picture one and picture two together varied from 40 to 60 to 80 milliseconds.
- The research question to be answered was to find proof for the hierarchical relation between the concept of position of the head and the wearing of glasses.

Why Calis always used pictures from the same set for the first and second picture, is a mystery to me. Perhaps it evolved from the research approaches of that time? These approaches put a heavy accent on contour similarities, so a material emphasis on help from the first picture to identify the second one. Certainly the material side pays a significant role but we also have to deal with the spiritual side of things and that means: knowledge, concepts, programs, strategies and so on.

To choose both pictures from one set makes the discussion and interpretation unnecessarily complex. A Gordian knot is created.
The question arises when the perceiver recognizes the identity of a person on the first picture and that picture is followed in the second picture by the same identity then you have a benefit from the first identification. And also when you have recognized the person on the first picture and the second person is a different one you can rule out one possibility and so enhancing the chance of a right identification.

This is all true, but Calis would argue that he corrected the results in these cases by means of chance correction.
Very good, one would say but for interpretability unnecessarily stressful.

My *first* proposal for modification of the approach is to use for the first and the second picture *different* persons. Never use as second presented pictures, pictures from the same set as the first presented pictures.
This solves all kind of nasty discussions about clarity of pictures, have benefit from the first picture in one way or the other, and so on: never will the perceiver be able on the bases of the first picture to establish the *identity* of the person on the second picture.

The only thing I can think of is that when we use as first picture always the husband or wife of the person on the second picture and these couples are famous. This phenomenon can easily be ruled out.

The only thing that should be of help is the established knowledge of the person on the first picture.
Or *not* be of help, because this knowledge is denied on the second picture.

What the perceiver brings to the identification of the person on the second picture is realized knowledge on the first picture and this realized knowledge (or enhanced perceptual orientation) can help the perceiver positively or negatively by the identification of the person on the second picture.

- Positively: when the person on the first picture was a woman and on the second picture there is also a woman then this knowledge established on the first picture is confirmed. Although it is another woman the expectation it could be a woman is already established and this is an advantage when there is little time left.
- Negatively: the perceiver has a disadvantage when on the first picture he has established that it is a woman and the second picture shows a man. The expectation is it a woman is denied and so the process has to start from the beginning in establishing new knowledge. When there is little time left it will be difficult to do so and as a result the correct identifications will drop.

The proposal for a *second* modification lays within the presentation time.
We know now that we have to make it more difficult for the perceiver to identify the second picture in order to get a differentiation in conditions. And this more difficult is pointed to the second picture because that's the one to be identified.
It's also more stable for the perceiver when he knows that the total presentation time in all cases is the same.

The proposal is to fixate the total presentation time for first and second picture together as 80 milliseconds and then within this total time to give the second picture more or less time.

For instance the first picture 50 milliseconds and the second picture 30 milliseconds in one condition and in the other condition the first picture 70 milliseconds and the second picture 10 milliseconds. The general expectation for the second condition is that the results will drop. This not yet answers the question whether there is a certain concept used or not.

Calis tried to improve the result by giving more time, but this improvement comes at the cost of an improvement of differentiation within conditions.

The result will drop when time is decreases that's a thing everybody can comprehend.

But when the perceiver can take something (knowledge) from the first picture to the second picture then within this lesser result, the result for the condition where the two concepts match will be relatively better than in the condition where they don't match.

A *third* proposal for modification is to investigate only one concept instead of two concepts. That's more easy to do and makes the interpretation clearer (more comprehensive).

So before the hierarchy hypothesis as stated by Calis there goes a simpler hypothesis and that is: in perception knowledge, a concept is used.

When more concepts are established by research then it becomes interesting to see how these concepts are related.

Summarizing the modifications:

- An absolute separation between first and second picture. This means pictures from the first set may never be used as second picture.
- Fixed presentation time for first and second picture together. And within this fixed time a sliding scale of presentation times for the first and second pictures separately.
- Limit the research to one concept only.

Research with modification proposals

I will try to describe this approach as if it was a manual for users in daily practice such as psychologists, pedagogues, police investigators in the field of perpetrator recognition and so on.

As a simple research example I choose to look whether the gender of a person plays a role in the identification of that person: is it a woman or is it a man? Let's call this the gender-concept. Why this concept? Perhaps it's an obvious choice to make and easy to investigate?

And why visual perception? Because the research of Calis was in this field and it is a seemingly easy accessible field.

Further on I will indicate that the approach in the visual field is the same as in other fields and combinations of fields.

Research question
The hypothesis, the presupposition of this research:

A perceiver uses a gender-concept in identifying persons.

> Who is this? -------Gender-concept -------- ! Name identification
> >> *time*

Training phase
Before the actual research there is a training phase. The perceiver is asked to identify two man and two woman. It's important that the person to be researched can identify these persons making no mistakes. We can therefor show these four pictures repeatedly and make sure that the same buttons with a corresponding name tag are pressed. Or when the perceiver cannot press a button can signal in another way which person he or she is seeing.

In other words the researcher must establish a *repeatable* correct answer.

This phase is often neglected, but is essential for a good result and essential for the width of the research: we do not want to investigate only "normal, healthy" people.

Task
The ultimate task during the investigation is the same as in the training phase: who is this? And then a choice from four of known persons.

The task is more difficult than in the training phase because the picture to be identified is *preceded* by another picture of another person.

The task stays: who have you seen? Pointing to the second picture.

A choice between four known persons.

The design of the investigation
The first set of pictures to be presented existed of 16 pictures of 8 man and 8 woman.

The second set of pictures to respond to existed of 4 persons: 2 man and 2 woman.

This is a minimal number. It must be possible to make a choice within each category.

In another research one can choose for more response possibilities.

The persons on the pictures in the first set were *not* members of persons in the second set.

Two time relations between first and second picture were used:

Time relation 1: first picture 50 milliseconds and second picture 30 ms.
Time relation 2: first picture 70 milliseconds and second picture 10 ms.

In all cases the total presentation time was 80 milliseconds.

When we combine the 4 persons from the second set with a man or a woman chosen from the first set we get 8 pairs of pictures.

Combined with 2 time relations we have 16 pairs of pictures.
So within a block of 16 pairs of pictures every person from the second set appears 4 times.
To prevent a systematic order of presentation we use random assignment without replacement.

In the two demonstrated researches we presented 5 blocks of 16 pairs of pictures. Why five? In principle one block is enough, but the research practice learns that the more data we get the better the more stable the end result.
The daily practice however would say the less time for an investigation the better.

Persons who participated with the research
Two female persons participated: R. and E.

Results[2]

	R.		E.	
	Time relation 1	Time relation 2	Time relation 1	Time relation 2
	50-30	70-10	50-30	70-10
First and second picture same gender	20	13	19	12
First and second picture different gender	19	4	20	4

Max score per cell is 20.

Graphically the results look like:

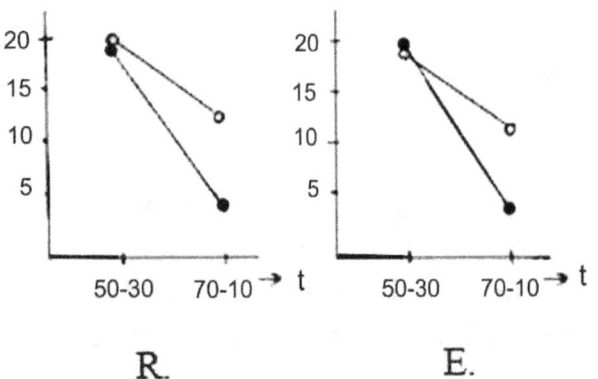

Both pictures of:
same gender (open balls)
different gender (closed balls)

R.   E.

General conclusion
We confirmed the approach of Calis (in a more limited sense) and now we can say that we have an instrument to determine whether a person uses a specific concept while perceiving something.

Generalization of the approach
We have now in the visual field confirmed that it is possible to determine whether a person uses a specific concept or not. This approach can be expanded to other modalities such as smell or sound.
We can also combine modalities e.g. a smell or sound followed by a picture.

---

[2] Look for the raw data on pages 27-29 and for the statistical analysis on page 30

With respect to the persons under research there are less restrictions: someone is blind but can be examined in the auditory filed or someone is deaf but can see and cannot speak then the researcher has to be creative to find something that can serve as a repeatable answer to some picture and different repeatable answers to different pictures.

How to research the hierarchy hypothesis?

Who is this ? -------- Gender-Concept------ ! Name identification
>>> time

Who is this ? ----- Age ------ Gender-Concept----------- ! Name identification
>>> time

The line of inquiry is as follows:
First confirm that both concepts are active within the participating person and then as a second step the relation between the concepts: is there a hierarchical relation? E.g. first the Age and then the Gender-concept? Or first the Gender-concept and then Age within this person?

Let's assume that Age comes first then our expectation would be:
When Age (A) and Gender (G) match this will lead to a higher score in time relation 2 (A+G+). In situations where Age match and Gender not (A+G-) the score will be lesser than (A+G+). Next where Age does not match and Gender match (A-G+). And finally the scores will be worst when both concepts don't match (A-G-).

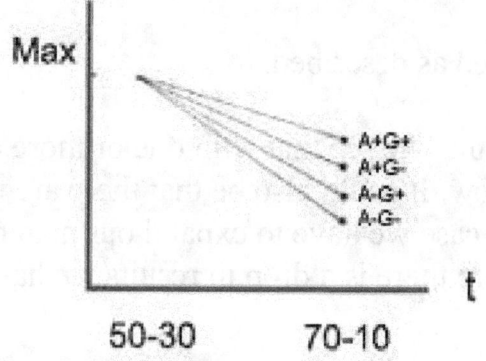

Hierarchical relation between Age (A) and Gender (G)

21

Unexpected extension

We can post the above described results as findings on the cognitive side of psychological research. This side delivers testable results, "hard facts". The reward for tackling the dragon in this field came from the other side of the coin: the emotion side.

In the cognitive field we can speak of an absolute separation of content between the first presented picture and the second one, on the emotion side there is no question of such a hard separation. Suppose the first picture is a woman and the second picture is Peter?
And the participant in the research knows both Peter and the woman as the ex-wife of Peter? The perceiver can have an advantage of this knowledge in identifying Peter.

These situations have to be excluded on the cognitive research side but on the emotion side these difficult situations can deliver indications about people with vague complaints and who cannot or dear to express themselves: what is bothering them?

Abuse

Suppose a girl was abused in her youth probably by someone in her own environment and this girl dears not to speak about this because there hangs a sanction, a taboo above the whole situation. It could also be that she is unconscious about the identity of the perpetrator. How can we solve this?

By means of the above describe method one can easily find clues:
The researcher collects photographic material from the direct environment of the girl, who is now a young woman: pictures of her mother, father, uncles, aunts friends, teachers and so on.
All these picture come in the set of first presented pictures.
Then a second set of pictures is composed with neutral persons for the girl with the vague problems.
The pictures are presented as described.

When we look at the results we see that with one or more combinations of pictures the results are low. It could also be that there are no significant differences. In the latter case we have to expand our material in the first set. But in the first case where there is a drop in results we have to study the pictures who caused this drop.

Anxiety blocks perception[3]

---

[3] Stranger at Killknock

It now depends on the expertise of the professional who treats the girl, how to handle this information. Remember we don't have hard facts here but we are speaking of clues. But clues can form a pattern and this pattern points in a certain direction.

Here we see a subtle cooperation between conscious and unconscious processes. From the conscious part is asked to identify a neutral and known person while the first presented picture can mobilize the unconscious part of the perceiver. Delivers the unconscious activity anxiety, then this anxiety affects the conscious activity without paralyzing the perceiver with fear.

How to let the unconscious speak in a conscious way.

Perpetrator recognition

This cooperation between the unconscious and the conscious we can also use in situations where we can speak of criminal accidents, like situations where someone is attacked or raped.

On television we often see a line-up of possible suspects and our victim has to pick the real perpetrator.

This is altogether not a good approach because when the accident was traumatic and lots of fear was mobilized, the victim when confronted with a group of possible suspects could be overwhelmed with fear (when the attacker is present) and point to another suspect in order to be released from this threatening fear. The victim wants to flee.

The above described method is much better in these cases. We can subtly activate the unconscious, the unconscious cannot make mistakes. In these cases it would be profitable that we also could make use of smells. When we are able to extract smells from possible suspects and present these smells by means of plastic tube as first "picture" and the victim recognizes this smell unconsciously than the results will drop dramatically when the attacker smell comes by. The researcher will have another strong clue.

Certainly in cases where it was too dark to see the perpetrator smells are a good alternative.

On schools

Also on schools this method can be useful.

Some students are bullied and threatened, but won't speak about it. This methods points to the persons that mobilize fear. How is the relation with the teacher? Also of interest in these days, misuses the teacher his powers? Is there fear in connection with this teacher? And why?

Now we have an instrument to investigate these questions and come up with clues. Clues that can form patterns and point in a certain direction.

With the growth of expertise of the professional using this method the professional can go deeper in the layers of personality, but as always this digging should be surrounded with responsibility and compassion.

Differences in research design between cognitive and emotion approach
The approach between cognitive and emotion research differs because in case of cognitive research we have to create conditions to determine whether a concept is used or not. This procedure is more restricted than the emotion approach which is more open and intuitive. In the cognitive approach we have systematically work through all the conditions to get an answer, in the emotion approach we have no conditions only pairs of pictures.

Statistical analysis
My professor of research methodology Dr. Peter Heymans always said: before you administer a statistical test look first at the raw data!
Are the raw data pointing in the expected direction?
If not, don't perform the test. Something you never forget.

In case of the reported research in general lines:
We speak of a main effect in the two time relation conditions:
Less time to perceive means worse results.
This is the physical side of the problem.
Now we want to know whether a concept is used, the spiritual side of things. We cannot see a concept. But we can notice whether a person uses a concept or not, it has it consequences in behavior. E.g. a teacher says to a class with students: on the left of the blackboard there it says... and so on. This teacher assumes that all the student have a kind of orientation concept that tells them this is the left side of the black board and that is the right side of the board. This seems simple but when a student does not have such a concept, he or she is likely to miss lots of information.
When we have to deal with visual stimuli there are two extremes: on the one hand a presentation time within which all stimuli can be identified and the other extreme is a presentation time within which no stimulus can be identified. In the latter case this presentation time is 0 milliseconds. The other extreme lies around 40 milliseconds dependent on the task (Calis, 1974 page 65).

In our experiment we see an almost 100% correct identification by 30 milliseconds with a picture of 50 milliseconds prior to the second picture that had to be identified, so even more difficult than a presentation of one picture with 30 milliseconds presentation time. In both extremes we see no differentiation in Genderorder so a female followed by a female or a male followed by a male against a male followed by a female or a female followed by a male this condition does not differentiate. This does not mean that there was not a gender concept used by the perceiver only we as researcher cannot demonstrate that this concept was used.

This is related to how the perceptual process works: Calis states that this process works from global to specific and is in a constant interaction with the world out there. This process seeks confirmation. When build up knowledge is confirmed the process can build further to more specific characteristics. When the process gets no confirmation it has to step back in the process to a point where confirmation was reached and so on.

So in the case that this process has established on the first photo that the person on it, is a male person and the second photo is also a male person but from a different identity, then this perceiver has benefit from this first identification because when the process has to step back the expectancy male is immediately confirmed in this case.

However when the first photo was a male followed by a second photo of a female, the process has to step back and tries to confirm the expectancy male but in this case this expectancy is denied and for a second time the process has to step back and has to confirm the expectancy female.

When there is very little time to do so (in our case 10 milliseconds) the Gendercondition same will benefit from the first identification and in the Gendercondition different the perceiver cannot benefit from the first identification and will be in a disadvantageous position.

This will lead to more positive identifications in the Gender same condition than in the Gender different condition, as is the case in our research: 13 to 4 correct identifications. For the second subject this was: 12 to 4. Both Chi Square scores significant.

So we can conclude that both subjects used a Gender concept in identifying the person on the second photo.

Basic literature

Calis, GJ J. (1974). Op het eerste gezicht. Onmiddellijke waarneming en gelaatsherkenning, dissertatie, Nijmegen 1974.

Calis, G. J., Sterenborg, J., & Maarse, F. (1984). Initial microgenetic steps in single-glance face recognition. Acta Psychologica, 55(3), 215-230.

Research data

The original research data on the basis of the proposed adjustments.
Two female students participated in this research: subject R. and subject E.

Structure of data-file:

First column = order of presentation
> There were 5 blocks of sixteen pairs. 80 trials in total. Each block
> contained 16 trials. Each trial within a block is a combination of a random
> pick from photographs of the first set and a random pick from
> photographs of the second set.
> The randomisation was without replacement.

Second column = combination of gender
 1 = female-female and male-male combinations of the first and second set
 2 = female-male and male-female combinations of the first and second set

Third column = time relation between first and second photograph
 1 = 50-30 msec
 2 = 70-10 msec

Forth column = choice from first set of photographs
 1 3 5 7 9 11 13 15 = male photograph
 2 4 6 8 10 12 14 16 = female photograph

Fifth column = choice from second set of photographs
 1 3 = male, maleA and maleB
 2 4 = female, femaleA and femaleB

Sixth column = the answer of the subject
 0 = no answer
 1 3 = male, maleA and maleB
 2 4 = female, femaleA and femaleB

 1 2 3 4 correspond with the photographs of the second set

Seventh column = status of the answer given by the subject
 + is a good answer, second photograph and answer are in line.
 - is a false answer.
 0 no answer. This answer is counted as a false answer.

Subject R.

1	1	2	11	1	1	+
2	1	2	14	4	2	-
3	2	1	9	4	4	+
4	2	2	2	3	3	+
5	1	1	6	2	2	+
6	1	1	8	4	4	+
7	2	1	5	2	2	+
8	2	1	16	3	3	+
9	1	1	3	1	1	+
10	1	2	10	2	2	+
11	1	2	15	3	3	+
12	1	1	7	3	3	+
13	2	2	12	1	2	-
14	2	1	4	1	1	+
15	2	2	1	4	3	-
16	2	2	13	2	1	-
17	1	2	9	3	3	+
18	1	2	8	2	2	+
19	2	1	2	1	1	+
20	1	2	11	1	1	+
21	2	1	3	2	2	+
22	2	2	16	3	2	-
23	2	1	12	3	3	+
24	1	2	6	4	4	+
25	1	1	13	3	3	+
26	1	1	1	1	1	+
27	2	1	5	4	4	+
28	2	2	15	2	2	+
29	1	1	10	2	2	+
30	2	2	7	4	3	-
31	2	2	14	1	2	-
32	1	1	4	4	4	+
33	2	2	13	4	3	-
34	2	1	7	2	2	+
35	2	1	16	1	1	+
36	1	1	8	2	2	+
37	1	2	9	1	2	-
38	1	1	11	1	1	+
39	1	1	4	4	4	+

Subject E.

1	1	2	11	1	1	+
2	1	2	14	4	2	-
3	2	1	9	4	4	+
4	2	2	2	3	3	+
5	1	1	6	2	2	+
6	1	1	8	4	4	+
7	2	1	5	2	2	+
8	2	1	16	3	3	+
9	1	1	3	1	1	+
10	1	2	10	2	4	-
11	1	2	15	3	3	+
12	1	1	7	3	3	+
13	2	2	12	1	4	-
14	2	1	4	1	1	+
15	2	2	1	4	0	-
16	2	2	13	2	3	-
17	1	2	9	3	0	-
18	1	2	8	2	4	-
19	2	1	2	1	1	+
20	1	2	11	1	1	+
21	2	1	3	2	2	+
22	2	2	16	3	3	+
23	2	1	12	3	3	+
24	1	2	6	4	4	+
25	1	1	13	3	3	+
26	1	1	1	1	1	+
27	2	1	5	4	4	+
28	2	2	15	2	1	-
29	1	1	10	2	2	+
30	2	2	7	4	3	-
31	2	2	14	1	2	-
32	1	1	4	4	4	+
33	2	2	13	4	3	-
34	2	1	7	2	2	+
35	2	1	16	1	1	+
36	1	1	8	2	2	+
37	1	2	9	1	1	+
38	1	1	11	1	1	+
39	1	1	4	4	2	-

40	1	1	5	3	3	+
41	2	1	10	3	3	+
42	2	1	3	4	4	+
43	2	2	1	2	2	+
44	1	2	15	3	3	+
45	1	2	2	2	2	+
46	2	2	12	1	4	-
47	2	2	14	3	2	-
48	1	2	6	4	4	+
49	2	2	5	4	2	-
50	2	1	12	1	1	+
51	1	1	9	1	1	+
52	1	1	10	2	2	+
53	2	1	1	2	2	+
54	1	1	15	3	3	+
55	1	2	2	4	2	-
56	1	2	11	3	2	-
57	2	1	3	4	4	+
58	2	2	13	2	3	-
59	2	2	14	3	2	-
60	1	2	7	1	4	-
61	1	2	8	2	2	+
62	2	1	4	3	1	-
63	2	2	6	1	4	-
64	1	1	16	4	4	+
65	1	2	7	3	3	+
66	1	2	12	2	2	+
67	2	2	11	2	2	+
68	2	2	6	1	4	-
69	2	1	15	4	4	+
70	2	2	3	4	1	-
71	2	1	10	3	3	+
72	2	2	16	3	1	-
73	1	1	9	3	3	+
74	1	1	1	1	1	+
75	2	1	14	1	1	+
76	1	1	8	2	2	+
77	1	2	2	4	2	-
78	2	1	5	2	2	+
79	1	1	4	4	4	+
80	1	2	13	1	3	-

40	1	1	5	3	3	+
41	2	1	10	3	3	+
42	2	1	3	4	4	+
43	2	2	1	2	2	+
44	1	2	15	3	1	-
45	1	2	2	2	2	+
46	2	2	12	1	1	+
47	2	2	14	3	2	-
48	1	2	6	4	4	+
49	2	2	5	4	2	-
50	2	1	2	1	1	+
51	1	1	9	1	1	+
52	1	1	10	2	2	+
53	2	1	1	2	2	+
54	1	1	15	3	3	+
55	1	2	2	4	4	+
56	1	2	11	3	3	+
57	2	1	3	4	4	+
58	2	2	13	2	0	-
59	2	2	14	3	0	-
60	1	2	7	1	1	+
61	1	2	8	2	2	+
62	2	1	4	3	3	+
63	2	2	6	1	4	-
64	1	1	16	4	4	+
65	1	2	7	3	1	-
66	1	2	12	2	2	+
67	2	2	11	2	1	-
68	2	2	6	1	0	-
69	2	1	15	4	4	+
70	2	2	3	4	2	-
71	2	1	10	3	3	+
72	2	2	16	3	0	-
73	1	1	9	3	3	+
74	1	1	1	1	1	+
75	2	1	14	1	1	+
76	1	1	8	2	2	+
77	1	2	2	4	3	-
78	2	1	5	2	2	+
79	1	1	4	4	4	+
80	1	2	13	1	3	-

Statistical analysis by SPSS V23

```
LOGISTIC REGRESSION VARIABLES VAR00007
 /METHOD=ENTER VAR00002 VAR00003
 /CONTRAST (VAR00002)=Indicator
 /CONTRAST (VAR00003)=Indicator
 /CRITERIA=PIN(.05) POUT(.10) ITERATE(20) CUT(.5).
```

Subject R.

Model Summary

Step	-2 Log likelihood	Cox & Snell R Square	Nagelkerke R Square
1	54,110ª	,420	,596

a. Estimation terminated at iteration number 6 because
parameter estimates changed by less than ,001.

Variables in the Equation

		B	S.E.	Wald	df	Sig.	Exp(B)
Step 1ª	VAR00002(1)	2,068	,716	8,337	1	,004	7,907
	VAR00003(1)	4,497	1,134	15,732	1	,000	89,762
	Constant	-1,423	,557	6,539	1	,011	,241

a. Variable(s) entered on step 1: VAR00002, VAR00003.

Subject E.

Model Summary

Step	-2 Log likelihood	Cox & Snell R Square	Nagelkerke R Square
1	58,438ª	,401	,563

a. Estimation terminated at iteration number 6 because
parameter estimates changed by less than ,001.

Variables in the Equation

		B	S.E.	Wald	df	Sig.	Exp(B)
Step 1ª	VAR00002(1)	1,395	,666	4,391	1	,036	4,037
	VAR00003(1)	4,336	1,101	15,524	1	,000	76,412
	Constant	-1,154	,513	5,058	1	,025	,315

a. Variable(s) entered on step 1: VAR00002, VAR00003.

CROSSTABS /TABLES=VAR00007 BY VAR00002 /FORMAT=AVALUE TABLES /STATISTICS=CHISQ /CELLS=COUNT PROP /COUNT ROUND CELL

Subject R.

Chi-Square Tests

	Value	df	Asymptotic Significance (2-sided)	Exact Sig. (2-sided)	Exact Sig. (1-sided)
Pearson Chi-Square	8,286[a]	1	,004		
Continuity Correction[b]	6,547	1	,011		
Likelihood Ratio	8,634	1	,003		
Fisher's Exact Test				,010	,005
Linear-by-Linear Association	8,079	1	,004		
N of Valid Cases	40				

a. 0 cells (0,0%) have expected count less than 5. The minimum expected count is 8,50.

b. Computed only for a 2x2 table

Subject E.

Chi-Square Tests

	Value	df	Asymptotic Significance (2-sided)	Exact Sig. (2-sided)	Exact Sig. (1-sided)
Pearson Chi-Square	6,667[a]	1	,010		
Continuity Correction[b]	5,104	1	,024		
Likelihood Ratio	6,904	1	,009		
Fisher's Exact Test				,022	,011
Linear-by-Linear Association	6,500	1	,011		
N of Valid Cases	40				

a. 0 cells (0,0%) have expected count less than 5. The minimum expected count is 8,00.

b. Computed only for a 2x2 table

Var0002 = column 2 = Genderorder
Var0003 = column 3 = Presentationtime relation
Var0007 = column 7 = Answers

Many thanks to Foeke van der Zee for his help with the data analysis.

Software

Presenting two pictures one after the other is easy to realize with current technical means. The computer monitors/screens have become super-fast, and the handling requires a program (software) that can offer the different pairs of photographs within pre-set time relations. The responses can be recorded and processed with results available at once. But are the screens reliable in the timing? I did some measurements on a laptopscreen but that was not convincing, so I made the effort to make a tachistoscope (a fast projector) myself. At the end of this paper in the introduction to the summary, some details of this tachistoscope making are given.

Working with different stimulation modes such as smells, sounds joined with photographs, more technical ability is needed.

While writing the computer programs for this approach I realized that we can make this paradigm even stronger by adding different pictures of one identity to the response set.

First part from Stranger at Killknock by Leonard Wibberley

Caitlin the Other House was leaning against the wall of the garden before her cottage when he met her and though he was prepared for some signs of change in her, he was surprised at the freshness and the color of her skin and the brightness of her eyes. Indeed it seemed that she had filled out and become very much younger and he was so taken aback on seeing this that for a moment he could find nothing to say to her.

"I see you are looking very well," he said eventually when he had recovered a measure of his wits.

"I thank you for saying it, but it is nothing less than the truth," said the woman. "And to save you your next question I will tell you that it is the stranger that has brought about this change in me."

"Indeed," said the doctor, "and how is it that he has been able to do that?"

"Who am I to explain the power that lies in the stranger?" said Caitlin the Other House. "Is there anyone in the world who can explain one person to another person? I'll tell you all that can be done. You can explain about another person only those things that you find in yourself. And any strange thing in another person you cannot explain at all but only wonder at.

Second part from Stranger at Killknock by Leonard Wibberley

"Do not put any faith in this stranger," he said. "You will only be deceiving yourself and when the truth comes it will be hard to bear. In all medicine there is no case of a woman of your age bearing a child. Even if you had a husband—"

He stopped, blushing, and vexed at himself that for all his medical experience, he could not take a clinical view of things and accept that it was possible for Caitlin the Other House to have a lover.

She laughed. It was a young laugh, full of teasing and pleasure. She stooped and picked up a pebble and threw it into the road. "Will you look for a moment at that pebble," she said.

The doctor looked at it.

"And now look at the mountain there beyond. The pebble is what you know and the mountain beyond is all the things that you do not know about. I will have my child soon. At the time of the walking of the stones, I will have it."

"At what?" asked the doctor.

"The time of the walking of the stones. They will come down to the lake soon to drink. They are thirsty and started already." She told him of the Man of

Stone and the Woman of Stone that had already moved behind the O'Flaherty cabin upon the mountain side.

"When did that happen?" he asked, interested despite his incredulity.

"The night before I met the stranger," she said.

"And I suppose that he had something to do with it?"

She did not answer and he left her.

Third part from Stranger at Killknock by Leonard Wibberley[5]

"Yes," said the stranger. He pointed to the mountain visible toward the horizon down the road.

'Tell me," he said, "is that Knockmaan?"

"Ah! We don't call it Knockmaan at all," said Caitlin. "Knockmor is what it is called."

"But the real name is Knockmaan. Isn't that right?"

"Well, it's not the name that we put on it," said Caitlin diplomatically, for she did not want to contradict a stranger and particularly such a very pleasant one. "I believe it was called Knockmaan in the old times, but now it is Knockmor."

"In the old times." repeated the stranger. "That was before the coming of Saint Patrick?"

"That is so, sir," said Caitlin. "You'll be interested in the plants on the mountain, perhaps. There's some of the strangest plants up there that you ever saw, so I'm told. There's plants that eats flies and some that eats moths and some that eats children—though that's just old talk. Still, there's a kind of a plant up there that if you stand on it, it will drain all the food out of you in a second and you'll be starving and trembling with hunger and if you don't eat that very minute you'll be dead the next one."

"Is that so?" said the stranger. "What's the name of that plant?"

"I don't know the English name for it. But the Irish name is fearnas na-n ocras. It means the plant of hunger."

"What does it look like?"

"There's the mystery of it," said Caitlin. "There isn't a person could tell you and sure you could not blame them, for with the fierce hunger that is on them when they step on the plant all they care about is eating whatever they can find so that they won't drop dead in a minute. Did it ever occur to you, sir, that it is only a man who has not been attacked by a tiger that can give you a good description of the beast?"

"I hadn't thought of it," said the stranger.

"Well now, 'tis true. For what kind of a description would you give if one came leaping out of the jungle straight at your throat when you weren't expecting it at all? Could you tell whether it had four legs or half a dozen, or whether the stripes on it ran lengthwise or along the body or whether indeed it was striped at all?"

"I see what you mean," said the stranger.

"Tis the same thing with the plant of hunger," said Caitlin. "Nobody that was a victim ever stopped long enough to look at it. And the rest of us, sure we

---

[5] In this part is shown what fear can do to perception.

35

wouldn't know it if we saw it in our stew. I suppose that you'll be going to the mountain looking for the queer plants that are on it?"

"Not especially," said the stranger. "Though I am interested in the mountain. But I have been struck by the truth of what you said about tigers."

"In what way were you struck by it, sir?"

'It applies to many things besides tigers. Enemies for instance. Many a man does not know his own enemy though others who are not involved can recognize him."

"It is a true word that you have spoken there," said Caitlin.

"And I suppose that there are many of us who if they found themselves in the presence of the devil would find him very likable and would not recognize him for what he was at all."

Author

Jan Sterenborg (1949) studied psychology from 1971 to 1978 at the Catholic University Nijmegen, now renamed to Radboud University.

While studying developmental psychology under prof. dr. Franz Mönks, the author met Dr. Ir. Frans Maarse and Dr. Gé Calis of the research department. The cooperation was very fruitful and led to a replication of the dissertation research of Gé Calis. Results of the repeated dissertation study are presented in Acta Psychologica (1984).

Before and during the study psychology, the author had contact with the Dutch sculptor Frans Coppelmans who had great influence in shaping the approach described here.

The Research Context
Laying the methodological basis for scientific psychological research

Preface

This article (1981) discusses the different research approaches.
A distinction between *descriptive* and *explanatory* research is coming forward.
This distinction is very important in our time as much descriptive research is interpreted as explanatory research. The outline of the research approach developed in the first part of this book is presented.
This text can be regarded as the methodological basis for scientific psychological research.

The Research Context

This paper originated after a conversation with Gé Calis about the Homunculus-problem. In this conversation Calis set the path to the following outline of the research context. We had this conversation in the corridor of the psychological laboratory in Nijmegen.

Abstract
In this paper the research-context is analysed in terms of the researcher's point of view and the subject's point of view.
The word Answer is used instead of the word Response in order to avoid simple associations with the S-R paradigm.

Introduction

Fig. 1 Task-situation.

Imagine a Perceiver (P) sitting behind a small desk and an Observer (0) watching the scene (fig. 1).
On the desk we see a bottle of wine (a), a packet of cigarettes (b), a glass(c), and an ashtray (d).

0: Can you tell me what there is on the table?
P: A bottle of wine, a glass, a packet of cigarettes and an ashtray. 0: Ok, is there anything else?
P: No.
0: Fine.

Now the observer changes the situation a little and blindfolds our perceiver (B). The following conversation ensues:

0: Tell me what there is on the table?
B: a,b,c,d.
0: Are you sure?
B: I think I am, but I will just check again.

Our perceiver puts her/his hands on the table and carefully reaches for the objects:

B: a,b,c,d.
0: Is there anything else?
B: I'll just have another check.
B: No.
0: Fine.

After a while the observer asks the same questions, receiving
in all probability the same answers (if provided, of course, the perceiver has
had the stamina and patience to continue).

0: Is there anything else on the table?
B: No.
0: Did you notice any changes?
B: No.
O: Fine.

This conversation can be repeated ad nauseam, until the perceiver is on the brink
of total boredom.

After a short break, the observer puts, unnoticed by the still blindfolded
perceiver, another object on the table. The blindfolded perceiver is then asked if
she/he is willing to continue with the experiment.

O: Tell me once again what there is on the table?
B: a,b,c,d.
0: Are you sure?
B: I'll just have another check B: a,b,c,d
0: So there is nothing else?
B: Certainly not ...oh just a second ...

At this moment the perceiver's fingers come into contact with an object that was
not there before. This gives rise to a certain amount of amazement and
excitement, and the conversation continues:

B: There is something else!

0: Tell me what it is.
B: I don't know: it feels like... is it e?
0: No.
B: is it z... no it is dh?
O: warm
B: ...is it h?
0: It is indeed. The experiment is now finished, and I would like to thank you for your co-operation.

Discussion of the task-situation

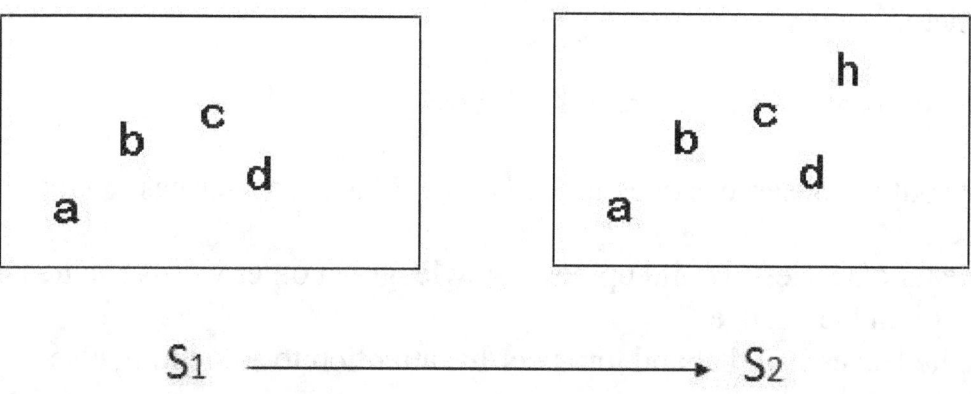

Fig. 2. Summary of the Task-situation.

In comparing the perceiver P and B situation we see that both ways of perceiving lead to the same result: i.e. knowledge of the situation S1 (fig. 2). After repeating the same procedure within the perceiver B-situation, we see that as soon as B is fully cognizant of the facts, there is a diminishing of attention (habituation). The situation is known, so no further checking is necessary. Attention can now be turned towards something else. This drop in attention happens within this constricted task-situation (enforced by the observer). If the observer were to play no active part in the discussion, our perceiver would in all probability leave the table and do something else. Let us now consider situation S2.

Our Perceiver's actions are based upon knowledge of situation S1. She/he expects the following: the bottle over there, the glass next to it and so on. Besides the four objects, there will be simply an empty table-top. Not having noticed the change that occurred from S1 to S2, our perceiver, after identifying (the four objects, fails to explore any further. At this moment the observer asks: "Is there anything else?" The blindfolded perceiver checks the rest of the table.

His/her expectation is: I will feel the table beside the objects. Suddenly our perceiver reaches the discovery:

No table!

A negative Answer to the Question: table here? Answer no!
Expectation violated.
Conclusion: something has changed.

From the observer's point of view we can say: s/he touches h, speaking in terms of the object h.
From the perceiver's point of view we can (must) describe the same situation as: no table here! Or old knowledge is not suitable any more. So we cannot speak of an object here.

These two points of view are essentially different!

If we leave out the observer's question: what is it? the perceiver has in fact 2 possibilities:
- either she/he tries to build up new knowledge, in other words she/he has to explain the change
- or s/he leaves it at that and turns her/his attention to something else (probably something known)

Let us concentrate on the first alternative. The perceiver is going to build up new knowledge. In doing so our perceiver is confronted with an infinite number of alternatives; it could be anything. The problem for the perceiver is to find a way in order to arrive at an answer that is in correspondence with h (fig 3.).

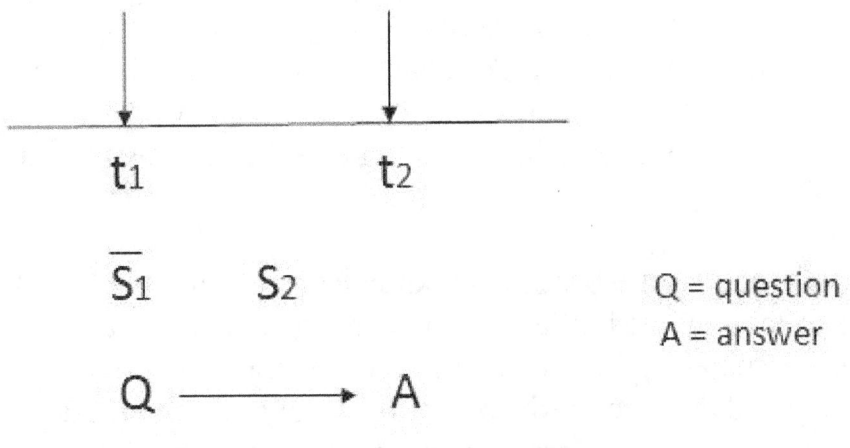

Fig. 3 Problem situation

Let us compare two alternatives:

- the perceiver gives as the first answer: h
- the perceiver gives a series of answers, some wrong, some containing correct elements and finally the right answer: h

```
?...........h
? ....e.......q...z...jh......jhhh..... h
```

Fig. 4. Two alternatives.

In the first situation we could say that the perceiver gives the Answer at-once (one-step), immediately.
In the second situation the Answer is reached by means of a number of intermediate steps (multi-step).

Are both situations really different?

Does the second situation explain for example more than the first one?

In the first situation there is a Question and an Answer. We have to guess what happened in-between (something must have happened if we think of the infinite number of alternatives). In fact this situation describes a problem to be solved: How to get from Question to Answer?

In the second situation we have a Question and a number of Answers. One could reason that by examining the in-between Answers, we could form an idea of how the perceiver manages to come from Question to the right Answer.
One idea could be that the information the perceiver realizes is in the beginning very global, becoming more and more specific with each step until 'the picture' is clear and the Answer can be given. We could say that we now have a more complex, a fuller description of the problem, but there is still the same problem to be solved! In fact the Question side shifts to the right (Answer) side, until we reach the same situation as described above (fig. 5).

```
?..........e.......q......................ih.........jhh..............h
Q-------A
        Q----A
             Q--------------A
                     Q--------A
                              Q----------A
```

Fig. 5. The Question-Answer shift.

47

Vice versa, if we look at the first Answer after the initial Question, we find ourselves in the same situation as described in the first alternative.
And similarly with all the in-between steps; they do not explain the problem they describe it.

We can define these as one-step or process-closed approaches.

Summarising briefly we could say:
In defining Question and Answer, we give in fact a description, not a solution of the problem.
In order to describe the solution, we must explain the step from Question to Answer. As we saw before, it is impossible to explain the step as a one-step process because this leads to irrelevancy. So the description of the solution has to be a multi-step description in order to explain it. This is its logical, necessary implication. We have to assume a multi-step process! In order to prevent us from the same mistake as discussed in the second alternative, we have to add restrictions to these steps in relation to Question and Answer. Before describing these restrictions, first some notational conventions and an example of the simplest form of a multi-step process (MSP) are given (fig. 6).

Question	Q
Answer	A
Step	Δ_i : i=1,N; N ≥ 1
Beginning of step (origin)	$B\Delta_i$
End of step (destination)	$E\Delta_i$

Process-closed conditions

OSP One-Step-Process
$$\begin{cases} \Delta_i : i=1 \\ B\Delta_i = Q \\ E\Delta_i = A \end{cases}$$

Process-open condition

MSP Multi-Step-Process
$$\begin{cases} \Delta_i : i=1,N; N \geq 2 \\ B\Delta_i = Q \\ E\Delta_i \neq Q \\ E\Delta_N = A \\ B\Delta_N \neq A \\ B\Delta_i = E\Delta_{i-1} \text{ (continuation rule)} \end{cases}$$

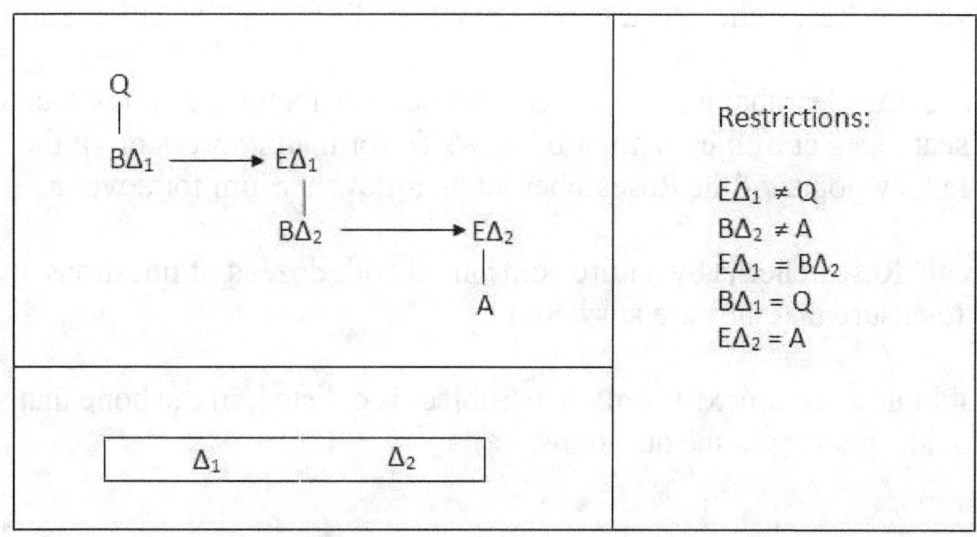

Fig. 6. Notational conventions and an example of a 2-step process, process-open condition.

So we can say that the steps between Question and Answer are related to Q and A but not directly related! There has to be something different from Q and A in-between.

In comparing the steps to each other, we can say they cannot be unrelated, because if so, we shift the problem to the last step and so we end up describing a one-step process! The notational consequence for this last statement is:

Delta i,j i=1,N-1; j=1,N; N>=2

By introducing the observer's question: what is it? We return to the old issue. So after a while the perceiver gives an answer. She/he has completed the task. How it was done we do not know yet, but we have given a short description of the problem. However, before discussing the solution, we must add another element. We can ask the following Question: what will happen if the answer of the perceiver and the observer are not in correspondence with each other? Here we meet the problem of inter-subjectivity or, put in other words, the problem of context or task interpretation. In the next part we will discuss this problem within the research-context.

Theoretical Implications

From now on, in order to keep in line with traditional conventions. Our observer will be called Researcher (R) and our Perceiver (P+B) will be called Subject (S).

It has become clear that a certain event can have different meanings according to the Researcher's or Subject's point of view. To form an impression of the possible viewpoints of the Researcher let us follow her/him for a while.

The "real" Researcher is by nature confronted with dozens of questions and the aim is to ensure that they are answered.

To avoid undue complexity, only one Subject is enlisted, in the hope that she/he can provide answers to the questions.

R
.

.
S

So we see here one Researcher with many Questions to be answered by means of the Subject.

Question-side	Researcher	Answer-side
??????	.	

Question-side	Subject	!!!!!! Answer-side
	.	

The Researcher wants to influence the Subject as little as possible so a context (related to the relevant Questions) is chosen.
The crucial danger here lies in the fact that the Researcher's context is not necessarily in correspondence with the Subject's context. If we increase the number of Subjects, situations are bound to occur where Subject defined contexts do not correspond to the contexts of the Researcher. In repeating this situation there can be shifts on two sides: the Researcher's Questions can change and/or the Subject's Questions can change, with the consequent change in Answers.

Having reached this situation, our Researcher thinks: at least we have to work within the same context, so the Question-side is consequently reduced into a single task which the Subject is asked to perform. (In the performing of the task by the Subject the Researcher concludes: same task interpretation).

Question-side	Researcher	Answer-side
?	.	

?		!!!!!!
Question-side	Subject	Answer-side
	.	

Our Researcher is intent on doing relevant research. So the Subject is given as much liberty as possible. The Subject has to perform the task: how the task is done the Researcher will decide afterwards, having analysed the data and having compared these results with explicit idea's formulated before. This situation seems fair enough. But still, if we compare different Subjects, some of them will act on the basis different task-completion criteria. The subject's will have different notions of the Researcher's expectations, which will be subjectively formulated, in order to arrive at a successful (in their own eyes) completion of the task (despite time-limitations or limitations of the set of response alternatives).

By repeating this situation, the possible shifts have decreased by one: if the Researcher asks the same Question, the Subjects can still act on the basis of totally different task-completion criteria, so the Question side is now closed but the Answer-side is still open. We can summarise both approaches as QA-open approaches.

Having noticed this problem, our Researcher concludes: in order to have complete task-interpretation on both sides (Researcher and Subject Side. The Question-side and the Answer-side must be identical between Subject and Researcher. And it must remain this way, even if the situation is repeated!

Question-side	Researcher	Answer-side
?	.	!
?	.	!
Question-side	Subject	Answer-side

So if the Researcher and the Subject ask themselves the same Question and give the same Answer, over and over again, the conclusion is reached: the situation is totally predictable.

The question which must now be asked is: What is there still to be known, if both sides (The Question and Answer side) are already known?
If both sides are known, I want to know how it is possible to get from one side to the other: How does it work?

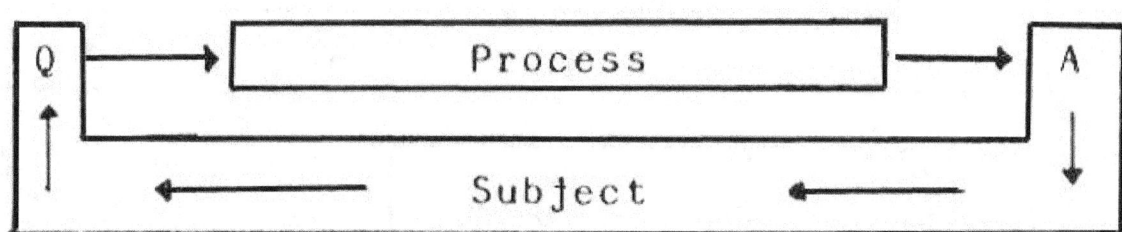

In fact, our Researcher assumes there is a process at work, and the job is now to explain that process, so explicit statements are formulated.

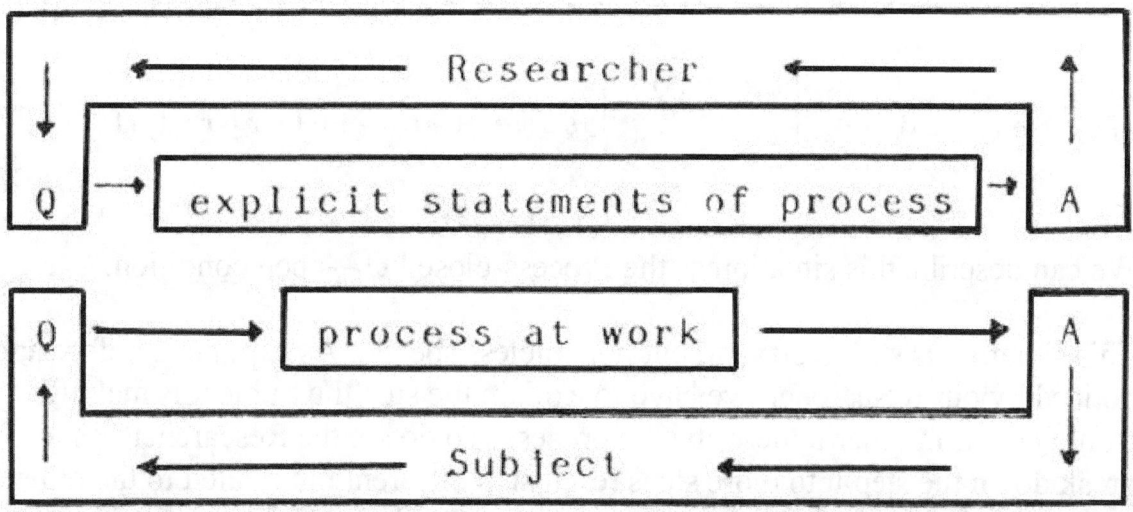

Now the Researcher formulates a suitable Question and Answer related to a certain Event {E0) and considers this as the operationalisation of the task the Subject has to perform.

The Answer of the Researcher is the expected or predicted Answer of the Subject. Further, the Researcher asks the Subject to perform the task and checks whether the expected Answer is in correspondence with the Answer of the Subject.

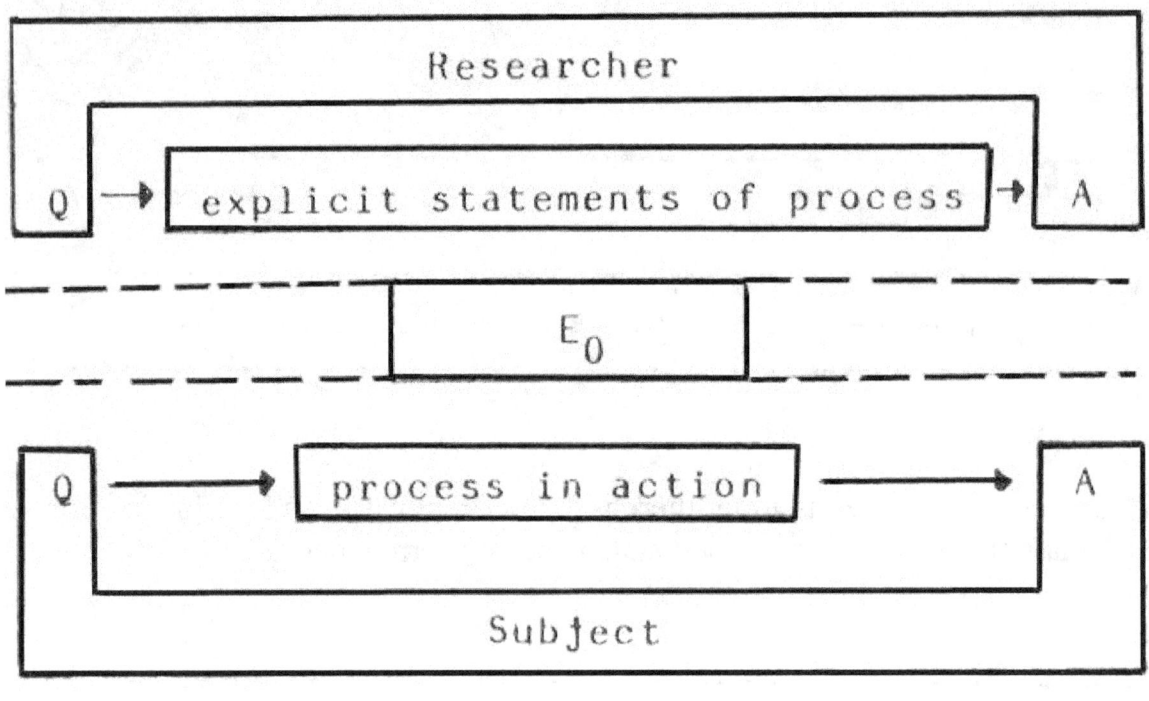

$$E_0 \text{ related to } \begin{vmatrix} QR = QS \\ AR = AS \end{vmatrix} \text{: same task-interpretation.}$$

We can describe this situation as the Process-closed QA-open condition.

This situation is still irrelevant, but nonetheless special. As explained before and noticed by our Researcher, we have to explain the step the Subject is making from Q to A, in order to describe the process. To do so, the Researcher has to break down the step into more steps, each step different but related to the other steps. The simplest case is 2 steps related to each other. In other words: the Subject, starting with question Q, is confronted with some event EO which will first activate step1 (in interaction with E1,1) and in relation to step1, activate step2(in interaction with E1,2), and give the answer A.

The next problem the Researcher has to solve is to compose an Event in Such a way that a differential A can be given.

The only way to get a predictable differential A is to decompose the Event in such a critical way that the change {born out of the decomposition of the Event), has a differential effect on the process in (inter)action on the Subject- side.

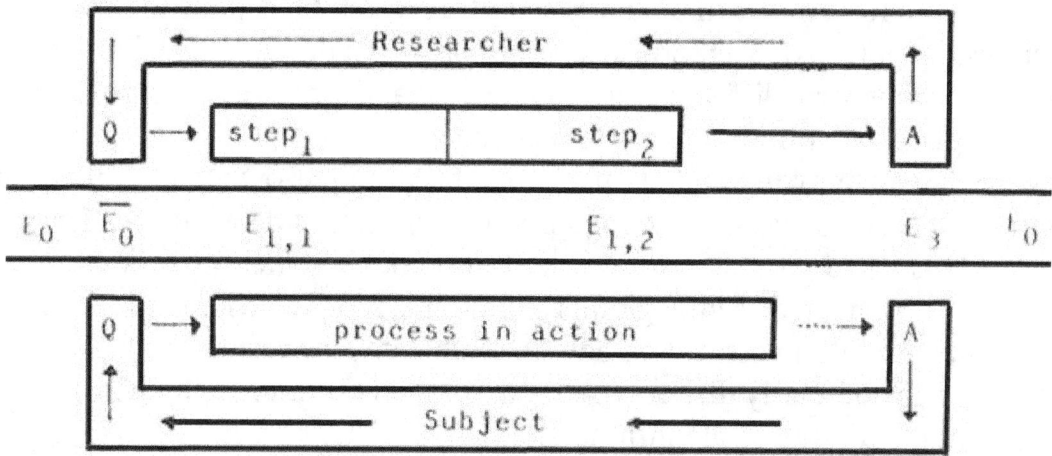

We can describe this situation as the Process-open QA-closed condition.

It goes without saying that El,1 and El,2 have to be similar with regard to step1, but they have to be dissimilar with regard to step2 in one case and have to be similar with regard to step1 and similar with regard to step2 in another case. From the Subjects side we can describe the situation as follows: starting from question Q and confronted with E1,1 the Subject concludes something and thus has gained knowledge. She/he tries to find further knowledge, because this situation is still too inconclusive for the answer to be given. An attempt is made, therefore, to obtain fresh knowledge which will lead to the correct answer.

Now confronted with E1,2 which was similar to E1,1 with respect to step1, our Subject can maintain this knowledge and reach in interaction with E1,2, a second conclusion. This conclusion enables the Answer A to be given. Confronted with E1,2 which is dissimilar with respect to step1 the gained knowledge cannot be maintained and the process has to start all over again. The probability of giving the right answer diminishes in this situation because there is too little time left.

In the first situation there was a possibility to build upon already established knowledge, so the probability of a right answer will rise. And so we see a differentiation between the two conditions when the predicted knowledge was in fact applied. When this knowledge is not applied we expect no differentiation.

Time-differences between Subjects, with respect to their Answer, are of no explanatory value. It is the quality of the Answer we are concerned with. This quality tells us whether we have made the right or wrong explicit statements regarding the process and the assumed knowledge within the Subject.

Concluding Remarks

Research can stress different aspects of the situation described above. One can leave the Subjects Question-side open, this means that within a certain context defined by the Researcher, the Subjects are free to ask Questions themselves, which results in the Answer-side also being open. To draw conclusions from these Answers is very difficult. One could say that they have descriptive value and no explanatory value. What fs to be done? We could restrict the Subjects Question side in allowing them to perform a task in their own way. This means that the Subjects Answer-side is open but the Question-side is closed. The Researcher has to define criteria in order to find out whether the task has been successfully completed. Although this situation is more restricted than the former, there is still no control on how the Subjects interpret the task. They can act on the basis of totally different task-completion criteria. Both approaches suffer from the disease of the infinite alternative.

Now if we restrict both sides, the Researchers and Subjects Question and Answer side, we are faced with a repeatable, controllable, predictable situation. When Subjects give the same Answers to the same Questions and they are in correspondence with the Researcher's Answers, we can say that the Subjects have the same task interpretation as the Researcher.

The whole research-problem now shifts from Question and Answer side, to the process-side of the problem. The Researchers question becomes: How does a subject arrive from Question to Answer? How does it work? The researcher has to explain this step.

Finding itself between a double pair of known-poles research must try to manipulate the process in such a way that we are able to predict the outcome of it. In order to do so, we need a situation that allows changes to be brought in, and this is the process-open QA-closed condition.

Now, if we do wish to influence the Answer-side within this approach there has to be a strong relation between the events $E1,1$ and $E1,2$. Both should consist for the most part of strong resemblances, but contain, on the other hand, crucial differences to be defined by the researcher. The crucial difference between both E's is directly related to the researcher's ideas of development of the process and the knowledge used in the process.

To summarize: the process proceeds in steps, and there is relatedness between the steps (organisation). In a more descriptive way: when the process reaches a particular stage in interaction with some event (this means limitation of answer possibilities, or higher probability of some answers being given), certain expectations are built up. It depends on the event in which way these expectations are confirmed or denied. If most of the built-up expectations are confirmed and a few are not, only a few corrections in the process have to be made in order to continue. The process-continuation stands in a very direct relation to knowledge that has been built up beforehand. If we make explicit what necessary steps the process has to make in order to reach the answer, we are able to predict the answer in terms of the differences of the changing event. The differences are operationalised in conditions beforehand. So different conditions produce different answer probabilities. If we find this, we can say that, from the researcher's point of view, knowledge has now been established about the working of the process and the knowledge used within the Subject.

Those theorists who put emphasis on ecological validity, arguing that within the laboratory situation we cannot obtain useful information, because this situation is too limited, do not see that it takes limitation in order to find (see!) something at all.

As indicated before, the step from Q to A is not restricted beforehand on the process-side!

Every subject is free to come in her/his own way to an answer.

There are of course situations where it is very difficult to create a task situation from which the researcher can draw the conclusion that the Subject has the same task-interpretation as the Researcher (Process-closed QA-open condition).

However, if the Researcher finds such a situation, from there she/he can turn to the process-open QA-closed condition to disclose the knowledge of the subject. The process-closed QA-open condition is a necessary condition for the Process-open QA-closed condition.

In this paper no differentiation has been made between the various modes of perceiving, e.g. visual, auditory, tactile, olfactory, etc. Because from a theoretical point of view, all these approaches are confronted with the same problem on the process-side! It is only a matter of the researcher's preference and skill whether one, two or more areas are selected.

In order to conduct research oriented towards the process-Side, it is necessary:
- to find a process-closed QA-open condition (repeatable task, same answers every time).
- make explicit statements about the process and the knowledge involved.
- create a process-open QA-closed condition in order to test the statements about the process and the knowledge involved.
- proceed.

Some observations

The big breakthrough with computers came when we turned from a linear approach such as DOS to circulatory systems as Windows and Linux.

In a linear system all functioning is top-down, a hierarchical approach: first, this process (= running program) then that process, and so on one by one, and one *after* the other.

Many processes run at the *same time* in a circular system. Lots of processes function independently, but other processes are interconnected. In the latter case the output from one process is the input for another process.

The overall system (Windows, Linux) regulates priorities. In a circular system the overall system looks into a pond, so to speak, and if somewhere a pebble is thrown in, the focus of the overall system, as part process of the overall process, goes to the peddle. All other processes just continue. The part process gets temporarily priority.

A circular process is many times more powerful than a linear process. A circular process requires better hardware: faster processor to control the many running processes in for the user a bearable time span and more memory.

Complementing the above story we can say that if an acting perceptual system is disturbed in its progress, it has to restart to check whether the hitherto accumulated knowledge is still adequate (orientation consistency). In that case the process can go into more specific knowledge building or follow a new focus.

Is the knowledge building at some point no longer adequate (denial of gained knowledge) then new knowledge should be built.

If the acting system fails to realize new knowledge then we can say that the system has "lost its way".

The observer must then find a point (by further stepping back in the process) from which new knowledge can be built. When he fails to do so then another observer can be of help (in severe cases s/he has to consult a therapist).

Acknowledgements

The author wants to thank Frans Duijf for his valuable guidance to improve the Dutch text with regard to the investigation.

Furthermore, the author would like to thank Chris van der Linden for his encouragement to give new attention to the old content.

John Siddle translated the Research context and provided me with critical questions.

Contents

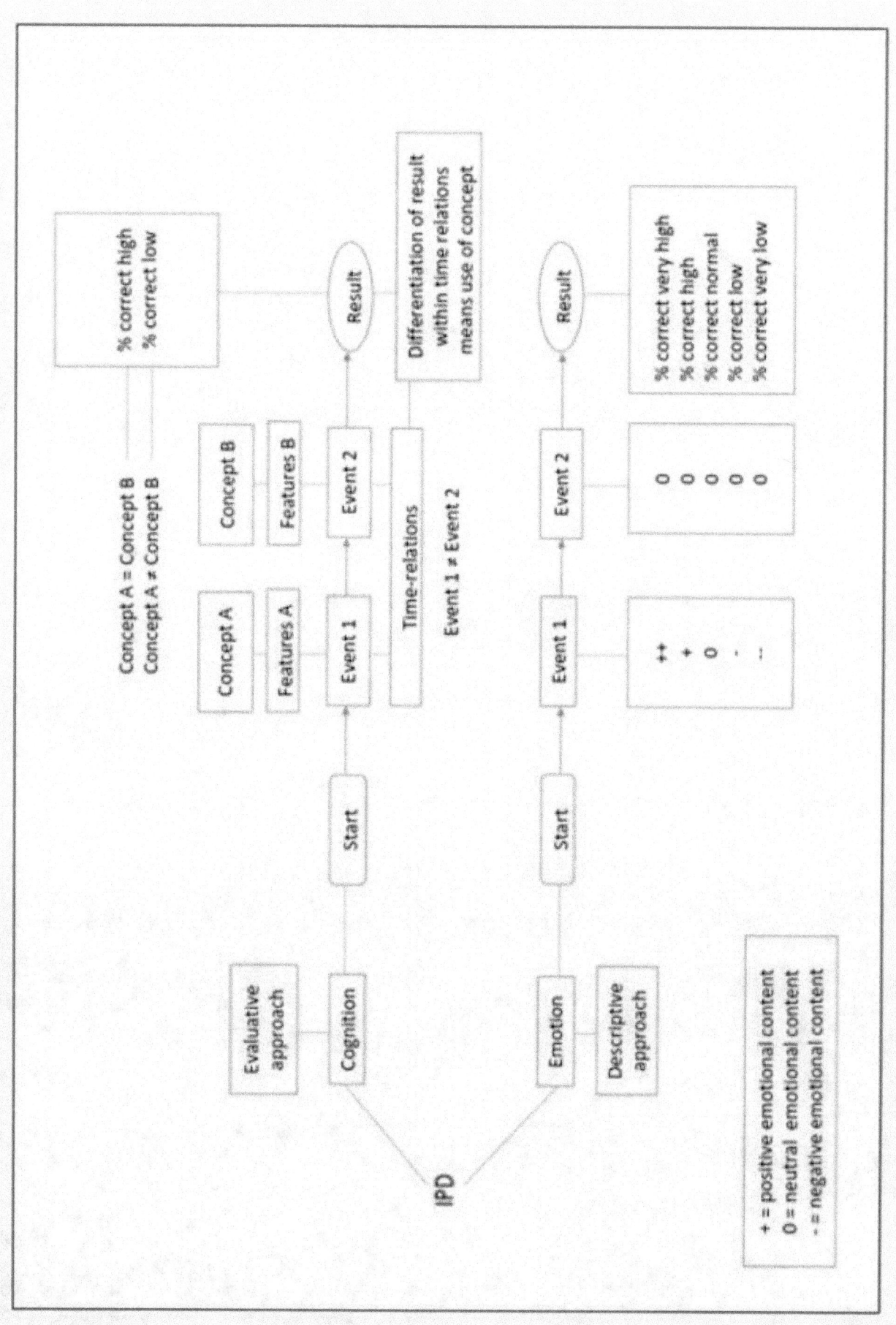

65

Introduction to summary

After elaborating the theory, some time was spent to build a tachistoscope myself because they are hard to get and they are also expensive.
In building a tachistoscope one gets more grip on the matter. The aim was to build an apparatus that could be connnected to a computer and by which pictures could be projected very fast. With the suggestions of the DIY projects on the internet of how to build a beamer plus the scientific articles, I managed to make a tachitoscope myself.
The reason I wanted to make a tachistoscope was practical.
What is a theory without the instrument to test it?
After building the tachistoscope, I discovered a more simple way to make one. With a simple and cheap LCD-LED-Beamer from China it was possible to make a tachistoscope by interrupting the power to the LED lamp.
The second problem was how can we switch OFF and ON the LED lamp very fast. This can be done by a SSR, a solid state relay, I used a Crydom D2D12. The SSR can be steered by a microcontroller (I used an Arduino Uno) and the microcontroller can be steered by a computer program that uses a USB port as COM port to which the Arduino is connected. The computer program was written in Microsoft Visual Basic.

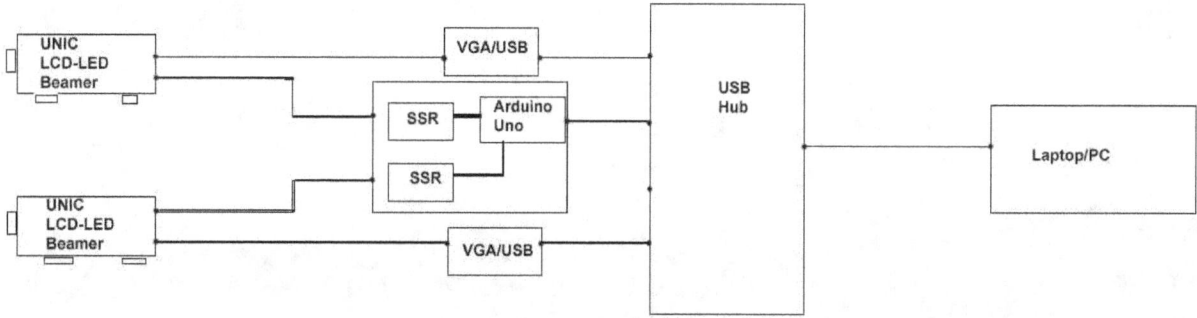

Diagram of a two channel tachistoscope

In preparing a simple anxiety test, I discovered that the most ideal configuration is a three channel tachistoscope. Two tachistoscopes are used to present the first picture and a third tachistoscope is used to present the picture that has to be identified.
In using two tachistoscopes for the first picture it is possible to give the first picture more life e.g. a small dog folled by the same dog but then larger. This gives the experience that the dog is coming towards the viewer. And if the viewer is afraid of dogs this "coming towards" is causing more anxiety than a static picture of the dog. This discovery completed my work on this new diagnostic instrument.

68

Technical literature

2010
Journal of Neuroscience Methods 187 (2010) 235-242

Multiple serial picture presentation with millisecond resolution using
a three-way LC-shutter-tachistoscope
Florian Ph.S. Fischmeister, Ulrich Leodolter, Christian Windischberger,
Christian H. Kasess, Veronika Schöpf, Ewald Moser, Herbert Bauer

2013
An LCD tachistoscope with submillisecond precision
Holger F. Sperdin & Marc Repnow & Michael H. Herzog & Theodor Landis

Published online: 7 February 2013 # Psychonomic Society, Inc. 2013

2015
Dynaflash The fast Japanes projector
http://www.k2.t.u-tokyo.ac.jp/vision/dynaflash/
Aug 4, 2015
http://tech.nikkeibp.co.jp/dm/english/NEWS_EN/20150804/430762/?P=1

Summary

Summarizing we can say that research with one tachistoscope, as Calis shows, leads to a description of phenomena in the visual domain.
The question was: what does a person see when pictures of famous people are presented to him or her in a very short way. The presentation time increased slowly with the next presentation (presentation times in milliseconds on the X-axis).

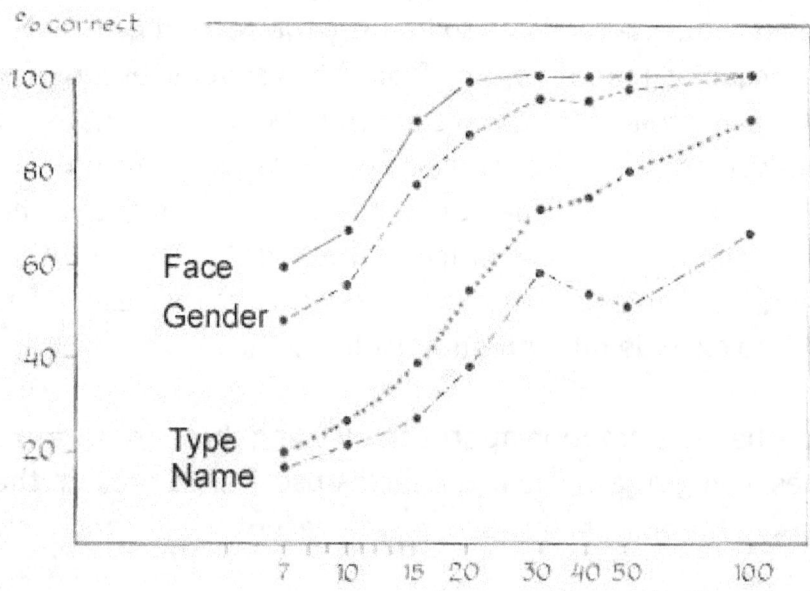

Figure1. Percentage of correct identifications for some "open-end" categories with unexpected portrait presentations in function of the presentation time.

As we can see in figure 1 we get an overview of what is easy perceptible and what is more difficult to see (Face, Gender, Type and most difficult the Name of the presented person).
This is determined, it does not explain why it is as it is.
The same applies in natural science if we want to test e.g. the hardness of solids, for this the scratch test has been developed that says that if you rub two different materials over each other then the softer material will get scratch marks. In this way we can draw up a hierarchy of hardness. A smart person can take advantage of this. If you do not have this knowledge, you will receive it through experience, trial and error.

With two tachistoscopes we can, as Calis shows, do research in the field of cognition. The perception of the position of a face takes place earlier (is a necessary condition for more differentiated perception) than determining whether someone is wearing glasses or not.

The theory explains the results: there is logic in the course of the perceptual process. This course goes from global to specific.

If, however, we carry out research of emotions with two tachistoscopes, we will lose explanation in the absolute sense, it will be describing again, but *interpretation* will be added. For example, we present a picture of a dog and then the person in question misses to identify the second picture. Then we interpret that result as fear of dogs or fear of this specific breed of dogs. It is in a certain sense an explanation, but we are not 100% sure of it, as is the case with cognition research. The fear can also be explained differently e.g. the presented dog is exactly the same dog as the neighbor's dog and the fear is linked to the neighbor and not so much to the dog. Apart from this, it can put the investigating psychologist on a track and it invites the researcher to do more research to exclude false presuppositions.

We can strengthen the emotion approach by using three (or more) tachistoscopes. Using two of the three tachistoscopes to present the first picture and using the third tachistoscope to present the second picture that has to be identified.

The advantage of two tachistoscopes for the first picture is that we can give that first picture more life: for example, we first show a small dog followed by the same dog but then a little larger: this gives the impression that the dog is getting closer and if there is fear for dogs, this approaching movement will evoke even more fear. In other words: an even worse result for the picture to be identified.

The same applies to people we fear: when they approach us, the fear increases.

So to be able to do all three types of research optimally, an arrangement with three tachistoscopes is ideal.

When this form of research becomes more widely known, perhaps more than three tachistoscopes will be needed to improve the research on emotions. I hope that the fast Japanese projector will become available (a 2d version of the fast projector is sufficient for this type of research).

We are talking about fear of dogs, but it will be even more interesting if we can use this approach to track down abusers of e.g. children.

For example, we can present photos of people around a child as the first photo and see by which photo or photos there is a lot of loss in the identification of the second photo. This of course does not mean that the individuals on the first photo who cause bad scores are by definition abusers, but it does give the psychologist data to investigate certain relationships more closely.

We can also get in this way a better picture of bullying behavior.

The same as with DNA determination: it is a good thing that there will be more awareness in the public domain that "you will not get away with it".

The phenomenological approach and cognition approach are important for a scientific foundation of psychology as a discipline.

The emotion approach is important for the practice of the individual psychologist. With this approach, every psychologist can further develop and shape his or her own field of expertise.

After 40 years of research and building on the work of Gé Calis and Frans Coppelmans, I hope that with all of this I have contributed to a solid foundation for a scientific and practical psychology.

Many people have supported me over the years, I am very grateful to them. Without this support this development path would not have been possible. In particular, I want to mention Chris van der Linden, Frans Duijf and Jerry Ent for their support and encouragement in the final phase of this project.

Arnhem, 12-2-2018
Jan Sterenborg

Calis, G.J.J. On the face of it. Immediate perception and face recognition. Thesis Nijmegen 1974.

A website: ipd-community.jouwweb.nl
Contact the author: website.vcr@gmail.com

ISBN   978-0-244-97145-8
NUR   770 776

© 2018,   Jan Sterenborg
            New resources for Individual Psychological Diagnosis
            Version 3.0
            Between Cognitions and Emotions
            The Research Context

www.ingramcontent.com/pod-product-compliance
Lightning Source LLC
Chambersburg PA
CBHW081101180526
45170CB00005B/1837